The Civil War in Indian Territory

Oklahoma Civil War Centennial Commission
J. Hardin Arias
　　The Civil War in Indian Territory

　　Bibliography

　　1. Indians of North America – Indian Territory – History – 19th century. 2. Indians of North America – Confederate States of America – Government Relations. I. Title

LC: E 78. I5
Dewey: 973.7

ISBN 978-1-935779-09-4

Oklahoma Civil War Centennial Commission

The Civil War in Indian Territory

J. Hardin Arias

Table of Contents

Foreword i

The Civil War in Indian Territory 1

Bibliography 19

Foreword

The Civil War in what is now Oklahoma, then known as "Indian Territory," has long been overshadowed by the larger, more dramatic battles and campaigns that occurred in the East. However, to the inhabitants of this land, the war brought devastation and tragedy that rivaled or even exceeded the experiences of other regions. The participation of the Five Civilized Tribes in the conflict stands as a poignant and complex chapter in American history, one that deserves deeper reflection and recognition.

This narrative unveils a story of divided loyalties, immense sacrifices, and the resilience of a people caught in a war not of their own making. The Five Civilized Tribes, uprooted from their ancestral homes and relocated to the wilderness by Southern influence, were faced with the harrowing choice of aligning with either the Union or the Confederacy. Despite their relative autonomy and independence from the political struggles of the United States, most tribes found their allegiances tied to the South, influenced by cultural ties, intermarriage, and the institution of slavery. Yet, there were exceptions — notably figures like the Creek leader Opothleyahola, who resisted Confederate pressures and remained steadfast in loyalty to the Union.

The war in Indian Territory was marked by skirmishes, battles, and campaigns that, while small in scale, held significant strategic consequences. The Territory's role in protecting the northern border of Texas and its impact on the Confederate war effort highlight its importance. The sacrifices made by Native leaders such as Stand Watie, the last Confederate general to surrender, and Black Beaver, who guided Union troops through perilous terrain at great personal cost, are testimonies to their leadership and fortitude.

This foreword aims to provide a lens through which to view the unique and often overlooked experiences of the Civil War in Indian Territory.

The accounts of survival, resistance, and adaptation against a backdrop of destruction and loss illuminate the enduring spirit of the tribes and the profound implications of this era on their future. By delving into this history, we honor the voices and stories that shaped this tumultuous period, ensuring they are remembered and understood for generations to come.

J. Hardin Arias

The part played by Oklahoma, then known as "The Indian Territory," in the Civil War, is not regarded as of great importance by many students. And the battles and campaigns involved were minor as compared to those in the East. As a result, the story of its participation is a most neglected field. But the war was as tragic, and even more so, to the dwellers in the "Territory" as it was to people in other parts of the country.

In all the annals of warfare, it is questionable if a more senseless happening has ever been recorded than the participation of the Five Civilized Indian Tribes in the American Civil War. To all intents and purposes, they were independent nations with nothing at stake in the White Man's embroilments. They had been uprooted from their ancestral homelands and transplanted in the wilderness by Southern people. Certainly, no allegiance was owed to the South. In what is now Eastern Oklahoma they had built a civilization which was virtually destroyed by the holocaust of a war that was not of their making. In the preceding quarter of a century, they had risen to a plane in education and economy which was comparable to much of the culture of the effete East and South.

Unfortunately for the Native Americans, most of their ties were with the South and Southern people. They had come from that area, many had intermarried with white people of the South, and a considerable number of them owned African American slaves. In addition, the agents representing the Federal Government in the Territory at the war's beginning were, in the main, from the South with strong sympathies for the Secession movement.

The decision of most of the members of the Five Civilized Tribes to go with the South had one definite and very important effect on the overall war situation. The Native Americans, aided by a few Texans, were able to keep Federal troops from reaching the Red River until the end of hostilities.

With her Northern border thus protected, Texas was enabled to send most of her troops and supplies to aid Confederate armies in the East.

News of the gradual withdrawal of the Southern states from the Union penetrated the Indian country bringing about much discussion and unrest. But not until the firing on Fort Sumter was the Territory galvanized into action. Federal garrisons were located at Fort Smith just across the border in Arkansas, at Forts Washita and Arbuckle in the Territory and at Fort Cobb in the country of the Plains Indians. Washington authorities realized Confederate troops from Arkansas and Texas would soon occupy these posts and capture garrisons and supplies if they were not speedily evacuated.

Lt. William W. Averell was sent from Washington in a hurry-up mission to warn Colonel W. H. Emory, commander of the border posts, to make his way with all troops under his command to Fort Leavenworth, Kansas, as best he could. The young officer reached Fort Smith just in time to inform Captain Samuel D. Sturgis, commanding this post, of his peril and then hurried on to search out Emory in the wilderness.

Within hours after Averell left Fort Smith, Captain Sturgis followed with his command, even as a steamboat loaded with Confederate troops was making its appearance. After a thrilling four day and night journey through hostile country, Averell reached the vicinity of Fort Washita. The approach of Texas troops had caused Emory to abandon this post and proceed toward Fort Arbuckle. On the march Averell, and, shortly after, Sturgis' command overtook the Union troops.

They arrived too late at Fort Arbuckle. It had already fallen, and the garrison had been captured. The same day members of the garrison who were paroled, joined Emory to hurry northwestward toward Fort Cobb, which Emory had also ordered abandoned. Thirty miles northeast of Fort Cobb the final junction of Union forces in the Indian country was affected. But what was to be done? Few had ever traveled northward over this area which since the days of Coronado had been regarded as "the Great American Desert."

The service of the renowned scout, Black Beaver, was now enlisted. This valiant warrior had settled into a prosperous agricultural life on the frontier. But he agreed to guide the troops to Leavenworth. The pursuing Confederates destroyed his farmstead and drove off his livestock. An unappreciative government never reimbursed the loyal Indian for the great sacrifice he made.

Jesse Chisholm, a half-Cherokee trader who remained loyal to the Union, accompanied the retreating troops. At what is now Wichita, Kansas, he dropped out of the caravan to remain during the war years, while Black Beaver and Emory's command continued to Fort Leavenworth. The evacuation of the Indian territory and the arduous trip over supposedly arid plains was accomplished without the loss of a man.

The Wichita and some smaller allied tribes of the Plains were in a fair way to adopting the White Man's way at the time war struck. But the wild tribesmen of the Southwestern Plains, Comanche, Kiowa, Arapaho, Cheyenne and Apache, interpreted the abandonment of Fort Cobb and Fort Arbuckle and the Texas posts as a sign of weakness and with redoubled fury continued raids on White settlements and Native Americans alike. The loyal Wichita fled to Kansas where they joined Jesse Chisholm for the duration, leaving a permanent recording of their Kansas sojourn by giving their name to Kansas' largest city. And, as a result of his flight with the troops over the route up which Texans were later to drive millions of cattle, Jesse Chisholm's name has been indelibly written into fame and story in the "Chisholm Trail."

Into the peaceful environment of the land of the Five Civilized Tribes in 1861 came a Yankee lawyer who was to render more harm to these Native Americans than have the machinations of all other White men before and since. Albert Pike was born in Boston. Possessing a wanderlust, he roamed over much of the frontier before settling at Little Rock, Arkansas. Upon the outbreak of the war, Jefferson Davis sent him as an emissary of the new Confederate government to the Five Civilized Tribes. Through judicious use of whiskey, flattery,

Creek Chief Opothleyahola

McKenny and Hall Lithograph, 1837

threats, cajolery, and rumored bribery, he persuaded virtually all of the Choctaws and Chickasaws, most of the Cherokees, and many of the Seminoles and the Creeks to side with the South. The end result of the Yankee lawyer's activities were four years of robbery, arson and murder in the Territory.

But he met his match in one Indian. Aging Creek chieftain Opothleyahola had fought against the removal of his tribe from their ancestral lands in Alabama. He remembered who had brought so much misery to his people. It was not Northerners but the same Southerners who were now endeavoring to destroy the Union he had come to respect. He would have no part of Albert Pike's blandishments. He strove desperately to maintain his tribe in a position of neutrality as did Chief John Ross of the Cherokees.

The defeat of General Nathaniel Lyons' Union Army at Wilson Creek in Missouri in August 1861, and the capture of Lexington and a Union Army by General Sterling Price in the next month, brought about a collapse of the Native Americans neutralist efforts. John Ross threw in the sponge and went with the Confederates. But not Opothleyahola. He remained loyal to the Union but, at the same time, realized the position of the loyal Creeks was rapidly becoming untenable. Preparations were made to go into exile in Kansas.

Colonel Douglas H. Cooper, a former Indian agent who went with the South, determined to prevent this exodus. Upon the approach of his command of Texans, augmented by Indian troops under command of Colonel Daniel Mcintosh, Lt. Colonel Chilly Mcintosh and Major John Jumper, the Creeks hurriedly left their homes and started for Kansas. They were overtaken and a sanguinary

battle fought at Round Mountain. The Creeks possessed enough ammunition to put up a real battle. Cooper's command withdrew after inflicting heavy losses on the Creeks but without accomplishing his purpose.

At Caving Banks, near Tulsey Town, the forerunner of modern Tulsa, Cooper again attacked Opothleyahola. Defection of a considerable number of his Cherokee Indian units caused Cooper to again withdraw from the battle. But a few days later, in the Battle of Chustenahlah, he accomplished his mission. The exhausted Creeks, with their ammunition depleted, could offer little resistance. After more heavy losses the survivors fled in wild disorder, abandoning their earthly possessions, and arriving in Montgomery County, Kansas, completely destitute.

Here they passed the first years of the war in utter poverty and here their heroic chieftain died. Opothleyahola proved himself to be one of the noblest of the Redmen. His name has never been given the recognition to which it is entitled.

The rest of the fighting during the Civil War by organized forces revolved around efforts of the Union troops to fight their way to the Red River. Because of the great distance from sources of supply, the North was never able to rally sufficient forces to push to Texas' northern border. Even worse, the constant bickering between Union commanders prevented any concerted and continuous campaign.

The first effort to invade the Indian country was organized in January 1862 by Brigadier-General James H. Lane, United States Senator from Kansas. His abortive expedition, which never got off the ground, was known as the "Lane Expedition." Lane got into an argument with the commander of the Department of Kansas, Major-General David Hunter, and on February 16, he resigned his commission.

Major-General Blunt assumed command of the Department of Kansas on May 15th. Field command of the expedition was given to Colonel William Weer, an attorney from Wyandotte, Kansas. The Confederates were attacked in the battles of Cowskin Prairie and Locust Grove. As was true of most of the battles in the Indian country, the Union command was left in possession of the field. Nearly every engagement followed the same pattern. The

Confederates fought well until ammunition and other supplies were depleted and then retired, but generally after inflicting such losses on the invaders that they were never able to follow up the victory gained. The first expedition was abandoned in mid-July after marching to within fifteen miles of Fort Gibson, apparently because of differences arising between the crusty leaders.

The second Federal invasion was undertaken by General James G. Blunt in October. At Fort Wayne he attacked Cooper's forces and drove them from the field. At the same time, loyal Native Americans attacked the Wichita agency near Fort Cobb. They destroyed the buildings of Agent Leeper and massacred most of the Tonkawa tribe which had sought refuge nearby.

In the following two months occurred rather inconclusive battles at Cane Hill and Prairie Grove. They did enable Colonel William A. Phillips to occupy Fort Gibson and destroy Fort Davis which had been established by the Confederates nearby. Dissension continued. Phillips could not get much needed reinforcements and supplies and withdrew to Kansas. The following spring Fort Gibson was again occupied by Union troops who thereafter held it precariously until the end of the war.

Much of the activities of the Federals from then on were devoted to holding supply lines open to Fort Gibson. The battles of Pea Ridge and Prairie Grove in Northwestern Arkansas broke the power of Confederates in that state. Fort Smith was taken and from it forays into Indian Territory were made. Indecisive battles were fought at Webbers Falls and Cabin Creek and in July 1863 General James G. Blunt left Fort Scott with the largest army to be gathered in the Territory. He arrived at Fort Gibson on the eleventh. Immediately the army crossed the Arkansas River and attacked General Cooper's command on Elk Creek in a battle which has become known as Honey Springs. It developed into the Territory's largest and bloodiest conflict. But, as usual, the Confederates exhausted their ammunition and retired to Fort Washita. Again, the Union force could not follow up its advantage and recrossed the Arkansas to Fort Gibson.

Cherokee General Stand Watie

Brigadier General, Confederate Army

(date unknown)

In August Blunt recrossed the river and an inconclusive skirmish transpired at Perryville. But here General Blunt's troubles with superiors came to a head and he was recalled in October 1863. The Confederate troops were demoralized, unpaid, poorly clothed, not always well fed, lacking arms, and ammunition and often poorly disciplined, but the constant dissension among Union officers prevented their taking advantage of such a situation.

Blunt was directed to report to Fort Scott. At Baxter Springs his command, consisting of two companies of cavalrymen, his band and a group of civilians were attacked by the notorious Quantrill's guerrilla band. The cavalrymen ignominiously fled leaving eighty-five helpless musicians and civilians to be massacred. Among the victims was James O'Neill, an artist for *Leslie's Weekly* who had accompanied General Blunt in his Indian Territory campaign. O'Neill drew the picture best known concerning the war in the Territory. It is a scene from the Battle of Honey Springs which was printed in *Leslie's Weekly* at the time.

The last serious campaign saw Colonel Phillips drive to within thirty miles of Fort Washita in February 1864. But valiant resistance on the part of Native Americans under the command of General Stand Watie, who had become recognized as the ablest of the Confederate Indian leaders, and Colonel Tandy Walker, caused this command, which made deepest penetration into the Territory, to retire. Stand Watie's activities earned him the sobriquet "The Indian Swamp Fox" because his warfare was modeled after the manner of Francis Marion of Revolutionary War fame.

The Indian country lapsed into such a state of anarchy that the law of the jungle was about all that prevailed. Stealing cattle and running them into Kansas became a prevailing activity.

Simultaneously with General Sterling Price's invasion of Missouri, which culminated in the Battle of Westport, Generals Gano and Watie moved up the Grand River Valley. At Cabin Creek in September 1864, they were so fortunate as to encounter what was probably the largest wagon train capture anywhere during the war. Loss of this huge number of supplies almost forced the Union to again abandon Fort Gibson. It did give the Confederates enough supplies of every kind to enable them to continue stiff resistance until the bitter end.

Defeat of Price at Westport and the retreat of his bedraggled troops through the Indian Country signaled the approaching end to Union and Confederate Native Americans alike. Union-sympathizing Native Americans began drifting back to their ruined homes as did their Confederate counterparts from Texas.

The last Confederate general to surrender was Stand Watie. On June 23, 1865, at Doaksville, he laid down his arms. The Native Americans were then subjected to the same severe reconstruction plan which was dispensed out by the Northern reformers who assumed control after Abraham Lincoln's death.

Bibliography: The Civil War in Indian Territory

1. **Abel, Annie Heloise.** *The American Indian as Participant in the Civil War*. University of Nebraska Press, 1992.

   A foundational study of Native American involvement in the Civil War, particularly focusing on their roles as soldiers and political participants.

2. **Debo, Angie.** *The Road to Disappearance: A History of the Creek Indians*. University of Oklahoma Press, 1941.

   Chronicles the history of the Creek Nation, including the devastating effects of the Civil War on their population and sovereignty.

3. **Edwards, Whit.** *The Prairie Was on Fire: Eyewitness Accounts of the Civil War in the Indian Territory*. Oklahoma Historical Society, 2001.

A collection of firsthand accounts offering perspectives on the Civil War from Native Americans and settlers in Indian Territory.

4. **Franks, Kenny A.** *The Civil War in Oklahoma*. Oklahoma Historical Society, 1980.

Provides an accessible overview of Civil War events and their significance in Indian Territory.

5. **Hauptman, Laurence M.** *Between Two Fires: American Indians in the Civil War.* The Free Press, 1995.

   Examines the complex roles and divided loyalties of Native Americans during the Civil War.

6. **Spencer, John D.** *The Battle of Honey Springs: The Civil War Comes to the Indian Territory.* Stackpole Books, 2006.

   Focuses on the largest Civil War battle in Indian Territory, highlighting its strategic importance and the diversity of the forces involved.

7. **Britton, Wiley.** "The Indian Regiments in the Civil War." *Chronicles of Oklahoma,* vol. 6, no. 3, 1928, pp. 248–267.

   Explores the organization, actions, and challenges faced by Native American regiments on both sides of the conflict.

8. **Gelo, Daniel J.** "Cultural Brokers in the Indian Territory: A Look at Confederate and Union Leadership Among Native Peoples." *Great Plains Quarterly*, vol. 18, no. 1, 1998, pp. 47–59.

Discusses the roles of tribal leaders as intermediaries and their navigation of allegiances during the Civil War.

9. **Starr, Emmet.** "Cherokee Indians and the Civil War." *Chronicles of Oklahoma*, vol. 8, no. 1, 1930, pp. 45–56.

Details the internal divisions within the Cherokee Nation and their impact on the war effort and post-war recovery.

10. **Confederate States of America.** *Indian Treaties (1861–1865).*

Official agreements between the Confederate government and Native American tribes, illustrating Confederate efforts to secure alliances in Indian Territory.

11. **U.S. War Department.** *The Official Records of the War of the Rebellion.*

A comprehensive collection of reports, orders, and correspondence, including detailed records of campaigns and operations in Indian Territory.

12. **Watie, Stand.** *Correspondence and Military Papers.*

The personal and official documents of Stand Watie, the only Native American brigadier general in the Confederate Army.

13. **Oklahoma Historical Society.** *Civil War in Indian Territory Digital Archives.*

A digital collection of photographs, maps, and documents related to the Civil War in Indian Territory.

14. **National Park Service.** *Honey Springs Battlefield and Visitor Center.*

Resources and interpretive materials focused on the Battle of Honey Springs, a key engagement in Indian Territory.

15. **University of Oklahoma Libraries.** *Indian-Pioneer Papers Collection, Western History Collections.*

Oral histories and narratives providing firsthand accounts of life in Indian Territory during the Civil War era.

www.ingramcontent.com/pod-product-compliance
Lightning Source LLC
Chambersburg PA
CBHW051435090426
42737CB00014B/2981